Through the Ages

ON THE COVER
Early Symbols for Basic Arithmetic

	**Niccolo Tartaglia**	1506–1555 Italian	Self-taught professor. Gave the addition sign.
	**Diophantus**	256 A.D. Greek	Father of algebra. Gave the minus sign.
	**Wilhelm Leibnitz**	1646–1716 German	Mathematician. Gave the multiplication sign.
	**Johann Rahn**	1622–1676 Swiss	Division sign first appeared in 1658.
	**Thomas Harriot**	1560–1621 English	Astronomer and mathematician. Helped Sir Walter Raleigh to survey Virginia and North Carolina. First to use symbols *greater than* and *less than*.

These mathematics symbols could be supplemented by many others for the practice was not uniform. The survival of arithmetical symbols has been due to the growth of the printing press and the development of mathematical theory. The evolution of these symbols has been slow throughout the ages.

Through the Ages

Virginia Moss Morgan

Illustrated by Raleigh Morgan, Jr.

VANTAGE PRESS
New York • Los Angeles

FIRST EDITION

Published by Vantage Press, Inc.
516 West 34th Street, New York, New York 10001

Manufactured in the United States of America
ISBN: 0-533-08646-9

Library of Congress Catalog Card No.: 89-90258

CONTENTS

PREFACE

How I hated mathematics!

As a student, I tried to escape its intrusion wherever and whenever I could. Then Sputnik was launched, and the new math was born. By that time I had become a teacher and been convinced that math must have some philosophical and practical meaning for students, especially. Sputnik spurred our pride, and the new math unfolded the *why* and *how* of numerals, a new challenge for students and teachers alike! The humanistic side of mathematics came alive when students interpreted and enjoyed depicting worldwide mathematical contributions on the school stage. Thus came about for readers, thinkers, and doers the book, *Through the Ages*, a bird's eye view of the history of mathematics.

This book discusses contributions to the various branches of mathematics and was written as a complement to the usual mathematics texts. It also gives the reader a deeper peek into the lives of those who gave so much to mathematics, whether they were mathematicians, philosophers, or scientists. Hopefully it will pique an interest to search for additional information and to appreciate multiethnic influences on the study of mathematics from ancient times to the present.

Just as students learn the fundamental processes of whole numbers and fractions, geometry and measurement, Roman numerals, statistics and probability, insight into the lives of contributors should be as exciting. The theories of Eratosthenes, Pascal, and Pythagoras are already well entrenched in the curriculum. Hopefully, this text may encourage curiosity for the designing of many more projects for the middle school student.

With great appreciation, acknowledgments are due to: my husband, Raleigh Morgan, Jr., for his patience, suggestions, and support; my colleague, Ms. Anna-Jane Meshkoff, librarian, for her unwavering encouragement; David Bragdon, DEC information consultant, for editing mathematical entries; and former sixth-grade student Carol McCarus, who inspired me to write this book. Finally, I must give my thanks and appreciation to the many students who questioned and then accepted a different approach to mathematics.

Through the Ages

1

INTRODUCTION

Have you ever wondered who devised methods for solving mathematical problems, or perhaps why mathematics is so valuable in our lives?

Mathematics has constantly developed from the time of counting sheep on fingers and toes to the orbiting of Uranus for data collection. Today we find that mathematicians still basically rely on the theories of multiethnic thinkers who lived hundreds of years ago.

Astronomy, the science of studying space beyond the atmosphere of the earth, was a favorite activity of ancient people. These early peoples developed their knowledge of the heavens by meticulously keeping a mathematical record of their observations. They also made a celestial calendar of events, which was compared to other star activities of the year.

As early as A.D. 150, via a simple sextant, people measured the position of new constellations. In addition, they knew the locations of many of the planets and other heavenly bodies by measuring the altitude of the sun in relation to the horizon. Ptolemy, the famous astronomer, believed the earth was the center of the universe. It was not until 1543 that the Polish astronomer Copernicus proved that the sun was the center of the solar system. Throughout the ages, man has traveled over land and sea by following the paths of the stars via a sextant. The forerunner of the modern sextant was not introduced until 1731. Today's latitude finders operate much like the ancient ones.

Because of this accumulated knowledge, today, in a more sophisticated sense, we can tell time, predict weather, assess home values, estimate population growth, locate oil reserves, and solve thousands of other problems, thanks to the gifts of many worldwide cultures.

A sextant, an astronomical instrument used for measuring angular distance between celestial objects.

Big and Little Dippers: these constellations are named for two star clusters.

Not all participants in this book are mathematicians, yet each person contributed an idea to improve mathematics. You will find that numerous facets of mathematics are described and illustrated. As you will notice, mathematical knowledge came in spurts of achievement, sometimes profuse, sometimes scanty. From ancient times through the fourteenth century, this irregularity was primarily due to the relatively late beginning of recorded history, wars, religious influences, and economic explosions.

In the fifteenth century, there was a new interest in the spread of mathematical pursuits. This shift was due to the growth of guilds, the mercantile industry, and the printing press. Mathematical ideas were even applied to improve the accuracy of artillery fire. This venture was first introduced by an Italian, Tartaglia.[1] By the middle of the fifteenth century, Greek and Arab mathematics were more easily available to European students.

The close of the sixteenth century saw the science of dynamics created. Dynamics is that branch of physics that studies the laws of force and motion. In fact there was a renaissance of interest in all branches of mathematics. The language of arithmetic and equations became fixed. The study of algebra and trigonometry advanced.

The seventeenth century saw great growth and embraced new mathematical fields to explore. Attention turned to human rights and the building of great universities. During the eighteenth and nineteenth centuries, there developed even more ideas, which were exchanged among interested countries. Now at the end of the twentieth century, mathematics has touched the depth of oceans as well as the heights of outer space.

The mathematical areas discussed are: arithmetic, algebra, trigonometry, calculus, probability, statistics, and computers.

Notes

1. Although artillery had been used as early as 1250, Tartaglia was the first person to apply mathematical knowledge to this science. He was born in 1499.

2 ARITHMETIC

Arithmetic is a language in which numbers and symbols are used to show relationships. Arithmetic serves as the basis for many branches of mathematics. It helps find short, simple methods of solving everyday problems. It counts, measures, and compares. In fact, counting is considered as the very first experience in arithmetic.

Let us begin our discoveries in the ancient world.

BABYLONIANS

These early people in Southeast Asia lived about 4000 B.C.–600 B.C. The Babylonians developed angle measurement and they knew that after dividing a circle (regardless of size) into equal parts, the sum of all angles equaled 360 degrees. The Babylonians used cuneiform (wedge-shaped) writing on clay tablets, many of which have perished. Their numbers were not deciphered until 1920. The Babylonians also introduced the "position number" system (an example could be 2 and 21). Notice that base 10 was universally used by most early mathematicians because humans were apt to use ten fingers or ten toes as a base for counting. Other groups used different bases. The Aztecs developed counting by grouping in twenties. The Babylonians used base 60 in measuring time (1 minute = 60 seconds, 1 hour = 60 minutes).

Babylonian cuneiform numerals on clay tablets. They were developed c. 3,500 B.C. and written with a stylus.

EGYPTIANS

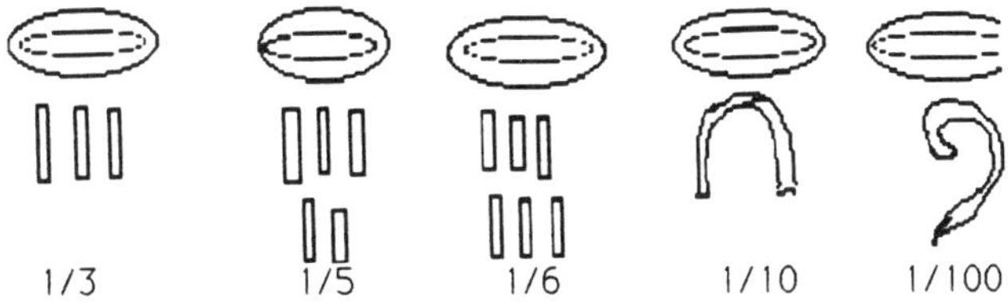

Early Egyptian expression of fractions.

By 2900 B.C., the Egyptians (north of the Sahara) had developed an advanced civilization. They gave us our first solar calendar, a number system, and an introduction to fractions. The Egyptians portrayed the concept of number by use of hieroglyphic symbols written on papyrus or stone. In fact, among the earliest surviving texts is the Rhind papyrus (1650 B.C.). Egyptian merchants probably brought our present numeration from Hindu trade routes; this knowledge was then disseminated throughout Europe before 1200.

Arithmetic played a great role in the average daily life for many Egyptians and their slaves as they worked on pyramids. These historic monuments were built with the use of a plummet, which helped cut bricks and stones accurately. Laid with little mortar, these "building blocks" fit closely within 1/500 of an inch. Egyptian mathematics was primarily suited for applications in agriculture and commerce.

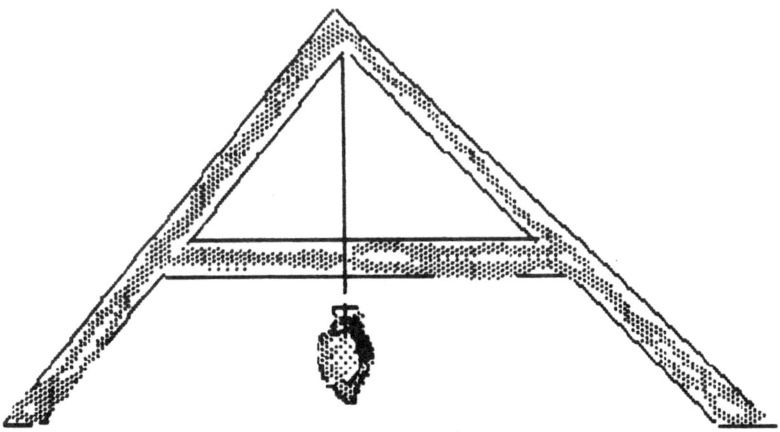

A plummet, an engineering tool for cutting stone.

Egyptians also surveyed land, divided the harvest, and measured grain. In fact mathematics was so valuable to Egyptian life that those who could afford it hired scribes to keep accounts. In addition, Egyptians knew the moon phases and that there were approximate 365 days in a solar year. *Area* also had meaning for them, e.g., $A = (1/2\ b \times h)$. To express fractions, the Egyptians used the sign for *mouth* placed above the denominator, according to Ifrah (1987:207).

Egyptian Numeral	Object Pictured	Decimal Numeral
	Tally	1
	Heelbone	10
	Coiled rope	100
	Lotus flower	1000
	Bent line	10000
	Fish	100000

Early Egyptian symbols or numerals.

CHINESE

China, considered a cradle of civilization, gave us the abacus, first used about A.D. 1200 and thus the earliest form of the

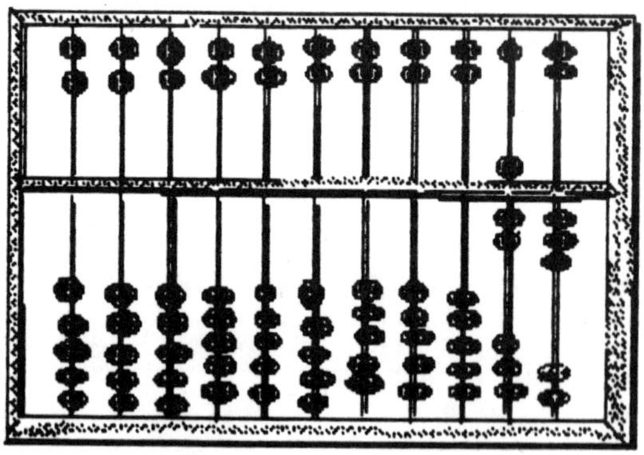

A Chinese Abacus

Different people have used many forms of abaci. The form illustrated consists of vertical rods, beads, and a horizontal bar. Each bead below the horizontal bar has a value of one and a value of five above the bar. Each rod has a given place value. The number illustrated is 73.

calculating machine. The abacus teaches place value, as well as addition and subtraction. They were acquainted with the square, the compass, the lever, and the ruler. They also knew about astronomical events happening in cycles. The Chinese devised a system of numeration that was first written on bark and bamboo. The Chinese recognized the use of integers. Black rods were used for negative values and red rods for positive values. Rod numerals developed differently in various countries, each independently.

BLACK AFRICANS

Fifteen hundred years before the colonial period (late 1300s) there were Great Kingdoms[1] also in black Africa such as Ghana, Mali, Kanem (bend of the Niger River), Songhai (savannah near the Niger River) and Benin (along the Guinea coast). These Great Kingdoms produced pottery, incense, spices, carved tusks, and introduced the pepper trade.

Meroë, in the Great Kingdom of Kush, was the outstanding center of an iron-smelting industry south of the Mediterranean

A bronze head of Benin.

coast. By the middle of the first century B.C., tools were fashioned into weapons. Meroë also showed evidences of basements in several ancient buildings, the luxury of indoor sanitation, the remains of the foundation of an astronomical observatory, wall drawings of a telescope[2] and the ruins of former splendors. There were also a series of wall drawings in numeric equations referring to astronomical events as well as the remains of pyramids and temples.

Benin, another thriving Great Kingdom city, gave the world the "lost wax" bronze technique for the casting of vases, jars, and sculptured heads. During the collection of taxes in Benin each person brought to the palace a cowrie shell for each animal owned. Separate piles of shells were made for sheep, another for goats, still another for cattle.[3]

By A.D. 900, black Africans had helped establish mathematical rates of exchange through gold, salt, and kola nut trade with Arabs across desert caravan routes. Brass weights for measuring wares were used. Each trader owned his personal set of about twelve weights of various sizes. A unified system of weights and measures was organized in Songhai (Ghana) for all markets and was strictly enforced by a network of inspectors.[4] Notice that black Africans used measured chemicals[5] for embalming purposes, and apportioned water for irrigating their fields.

HINDUS (East Indians)

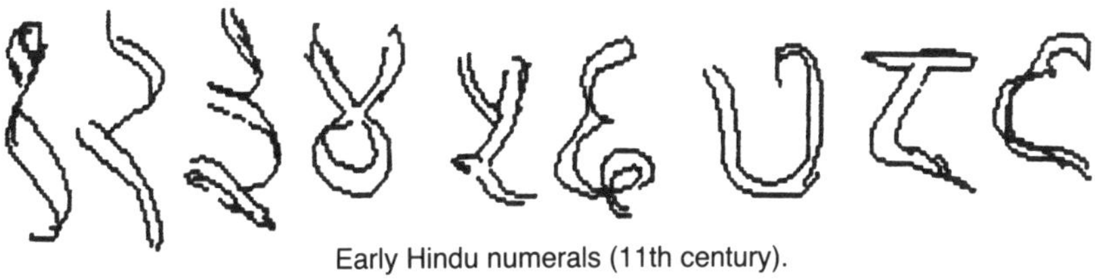

Early Hindu numerals (11th century).

In India, the early use of mathematics was related to religious rituals, such as the construction of altars and temples. The Hindus gave us "place value" of a numeral and the use of zero, which is a place holder. Zero means "empty" and may have been invented as early at 100 B.C. We learn that numerical signs used by the Hindus had variable values depending on their position. Signs for the numbers 1 to 9 had a place value before the eighth century A.D. Ifrah[6] tells us an anecdote about the origin of number place value.

One day, Sessa, a Hindu scholar and inventor of chess, presented this game to the then king of India. He was so fascinated by its many ingenious challenges to the player that he sent for Sessa, so that he could personally reward him for his extraordinary creation.

> "For your remarkable invention," the king said enthusiastically, "I am going to give you any reward you want. To you my generosity will be boundless."
>
> "I'm deeply grateful to you, Your Majesty," replied Sessa, "but I won't ask for a great reward. I'd simply like you to give me as many grains of wheat as it would take to fill the sixty-four squares of my chessboard if the first square held one grain, the second two, the third one four, the fourth one eight, and so on, doubling the number of grains from one square to the next."
>
> "Your request is much too modest!" exclaimed the king, surprised. "You offend me by asking for something so unworthy of my benevolence and so insignificant compared with the magnificence of what I can give you." But Sessa insisted till the king resigned himself and ordered his vizier to bring the "bag of wheat" that Sessa wanted.
>
> A few days later, the king asked if that fool Sessa had

taken possession of his meager reward. "The reckoners attached to your court haven't yet determined the numbers of grains," said the vizier. "They hope to finish tonight, before dawn." But the next morning the royal reckoners had not yet calculated the amount of wheat that had to be given to Sessa.

The king was exasperated.

"How can they take so long to solve such a simple problem? I order them to finish before tomorrow morning."

The problem was still unsolved the next day. Judging them to be incompetent, the king angrily dismissed his reckoners and replaced them with others. But the new ones made no faster progress than the old ones. After several days of ceaseless work, the leader of the reckoners came to the king to announce the final result.

"Well, have you given Sesssa his reward?" the king asked impatiently.

"Forgive me for saying this, Your Majesty, but in spite of all your wealth and power it's impossible for you to provide that reward. Even if you emptied all the granaries in the kingdom, you'd have only a minute fraction of the amount of wheat you promised to Sessa. There's not that much wheat in the granaries of all the kingdoms on earth."

"Be more specific! Exactly how much is it?"

"Your Majesty, the number of grains of wheat that Sessa wants is eighteen quintillion, four hundred and forty-six quadrillion, seven hundred and forty-four trillion, seventy-three billion, seven hundred and nine million, five hundred and fifty-one thousand, six hundred and fifteen."

The number can be written in our modern notation as: 18,446,744,073,709,551,615.

Bhaskara, a famous Hindu astrologer, was also a mathematician of exceptional ability. We know he was born about 1114. One of his books, *Lilivaki*, gives rules for all mathematical functions. Bhaskara wrote a book of problems for his daughter. Following is an excerpt:[7]

> A snake's hole is at the foot of a pillar which is 15 cubits high and a peacock is perched on its summit. Seeing a snake at a distance of thrice the pillar's, gliding toward his hole, he pounces obliquely upon him. Say quickly at how many cubits from the snake's hole do they meet, both proceeding an equal distance?

MAYAN

Mayan temple-topped pyramid.

At the height of a remarkable civilization from 300 to 800, the Mayan Indians invented a calendar, built temple-topped pyramids, and invented a system of numbering based on the digit "0." This system was written in a series of dots and dashes (see illustration below). They also made a table of dates that foretold the eclipse of the sun. Their records used a system of knotted strings called a "quipu."

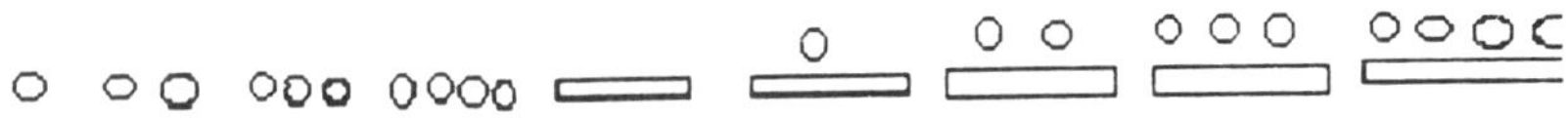

Early Mayan numerals.

GREEKS

The ancient Greeks learned many mathematical techniques from the Babylonians and improved upon them. They loved logic and associated mathematics with philosophical principles. The Greeks were not just satisfied with a result. They wanted to understand "why" with proof. Engaged in formally investigating mathematical laws, the ancient Greeks gave the world "The Sieve of Eratosthenes,"[8] The Pythagorean Theory," and numerical ideas for musical harmony. The Greeks also gave us "palindromes," a group of numbers (1881) or words (Able was I ere I saw Elba) that read the same backward as well as forward. For mathematical symbols, the 24 letters of the Greek alphabet and three letters borrowed from an earlier alphabet were used.[9] The Greeks were the first to treat mathematics as an art, geometrical figures creating different forms.

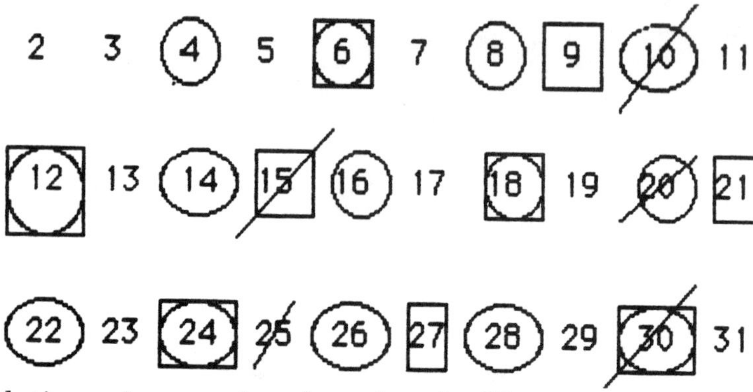

Isolating prime numbers by using the "Sieve of Eratosthenes."

ROMANS

The Romans gave us still another system of enumeration, Roman numerals. They used their mathematical knowledge primarily to build roads with drainage and houses containing a central heating system. The Romans developed mining techniques, used a relay system of postal service, and built arenas that seated as many as fifty thousand spectators. The Romans knew the strength of the arch, which they designed for highways and aqueducts of the empire.

Roman numerals today are useful for naming chapters, reading clock dials, and dating monuments. We should remember also that we use letters as numbers very extensively in another branch of mathematics—algebra.

I II III IV V

VI VII VIII IX X

L C D M

Roman numerals.

Instead of creating new symbols for power of ten greater than 1,000, a bar was placed over an existing symbol, e.g., $\bar{L} = 1{,}000 \times 50$. The Roman system was both additive and subtractive.

Notes

1. Basil Davidson, *The Lost Cities of Africa* (New York: Little Brown, 1959).
2. Cheik Anita Diop, *L'afrique noire pré-coloniale* (Paris: Présence Africaine, 1960).
3. Ibid.
4. Claudia Zaskavskia, *Africa Counts* (Boston, Ma.: Prindle, Weber and Schmidt, Inc., 1973), 27.
5. Carter Woodson, *The African Background Outline* (New York: New American Library, 1969).
6. G. Ifrah, *From One to Zero* (New York: Viking Penguin, 1987), 434–37.
7. Howard Eves, *An Introduction to the History of Mathematics*, 4th ed. (New York: Holt, Rinehart, Winston, 1938), 181.
8. A method for finding prime numbers.
9. Francis Mueller, *Arithmetic* (New York: Prentice Hall, 1956), 63.

3

ALGEBRA

Algebra had its beginnings in arithmetic and developed a more complex nature as time went by. It is a branch of mathematics where unknown numbers are represented by letters. In solving problems, one must supply the value of the letter. The algebraic equation may be added, subtracted, multiplied, or divided. By the use of symbols, a problem can be expressed in a brief form known as a formula.

As trade was stimulated and geographic discoveries made, many economic and political changes occurred. Interest in mathematics quickened.

DIOPHANTUS (Greek)

This mathematics personality lived about A.D. 250 and used symbols of his own for the unknown. He wrote *Arithmetic*, a thirteen-volume book containing about 150 problems, each illustrated and solved in an ingenious manner. He selected the sign "-" meaning "to subtract." His method was to present a problem and then solve it by changing the problem using "x" as the unknown. The numerical solution was then analyzed and the result was considered as proof. An example from Ball (1960:107) follows:

> (i) Find four numbers, the sum of every arrangement three at a time being given;
> say 22, 24, 27, and 20 (book I, prob. 17).
> Let x be the sum of all four numbers; hence the numbers are x - 22, x - 27, and x - 20.
> therefore x = (x - 22) + (x - 24) + (x - 27) + (x - 20).
> therefore x = 31.
> the numbers are 9,7,4, and 11.

Diophantus, a Greek algebraist.

MUHAMMAD IBN MUSA AL-KHOWARIZMI (Arab)

Astronomy was the most popular study in the Arab world. Because of this fascination with the heavens, theories of algebraic computations developed rapidly and Arab scholars were warmly welcomed by universities of the then-known world as teachers. Arabs further developed positive and negative numbers and worked with common fractions. They even designed a method of extracting equal factors called *root*.[1] The word *root* is developed from an Arabic concept meaning "source" or "essential." They thought of a root sign as being analogous to a plant root and so the root sign[2] was created.

Al-Khowarizmi (about 825) improved algebraic knowledge and solved equations by using rules. Our word *algebra* comes from the Arabic *al-jabr*, which means "transposing."

Carcassonne, a walled town in Southern France where Arabs were popular and welcomed as scholars.

OMAR KHAYYÁM (Persian)

From "The Rubáiyát of Omar Khayyám"[3]

Awake! for Morning in the Bowl of Night
Has flung the Stone that puts the Stars to Flight;
And Lo! the Hunter of the East has caught
The Sultan's Turret in a Noose of Light.

●●●

A Book of Verses underneath the Bough,
A jug of Wine, a Loaf of Bread—and Thou
Beside me singing in the Wilderness—
Oh, Wilderness were Paradise enow!

—Translated by Edward Fitzgerald

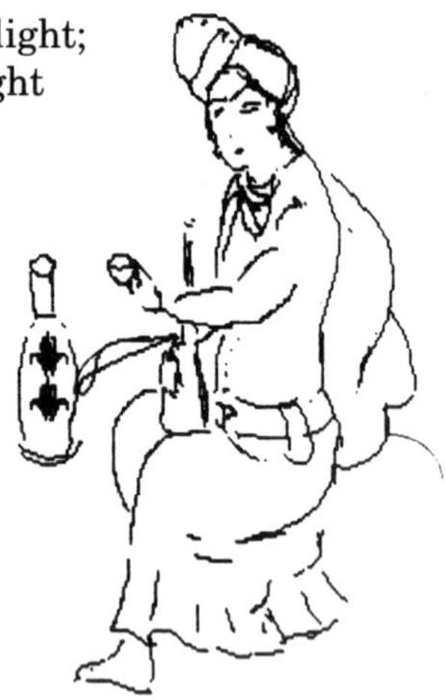

The Persian poet, astronomer, and mathematician, Omar Khayyam.

Omar Khayyám was a Persian poet, astronomer, and mathematician (1050–1123) who wrote a book giving his views on algebra. He also devised a Moslem calendar.

LEWIS DODGSON (LEWIS CARROLL, English)

This Oxford mathematician, who used his story *Alice in Wonderland* (1865) to amuse children and poke fun at human frailties, was a quiet mathematics lecturer and photographer of children. He wrote many books and papers concerning mathematics. His first work was *A Syllabus of Plane Algebraical Geometry*. Carroll used mathematical references for amusement in his stories,[4] e.g., "I'll take measurements," said the Queen (p. 179) " . . . at the two-yard peg . . . "(p. 180); "and if you take one from three hundred sixty-five what remains?"(p. 257); "I wonder

Alice and the White Rabbit, two characters from Lewis Carroll's *Alice in Wonderland.* Lewis Carroll (Lewis Dodgson) was a mathematician and a graduate of Oxford.

how many miles I've fallen"(p. 3); "I wonder what latitude or longitude, I've got to?"(p. 3); " . . . she was only ten inches high . . . "(p. 8); " . . . the multiplication table doesn't signify . . . "(p. 15); " . . . the earth takes twenty-four hours to turn around on its axis" (p.62); " . . . the different branches of arithmetic—Ambition, Distraction, Uglification and Derision" (p. 186).

Notes

1. The square root of a number is a second number whose product with itself gives the original number, e.g., $\sqrt{16} = 4$, $\sqrt{49} = 7$, $\sqrt{100} = 10$. The square root of 64 written in measured rhyme 8.
2. See the end of footnote above.
3. *The Rubaiyat* is a collection of four-line stanzas is primarily about nature written in measured rhyme. It was translated by Edward Fitzgerald in 1859.
4. See for example his *Alice in Wonderland* and *Through the* the *Looking Glass*.

4

GEOMETRY

Geometry had its origin in surveying the land and became a science. In geometry, one finds information about figures of various shapes. All of nature[1] has myriads of different geometrical shapes. Our feelings respond to these shapes. For instance, horizontal lines and curved lines give a feeling of serenity, whereas jagged lines give a feeling of restlessness or high energetic activity.

One fascinating outgrowth of geometry is topology. Topology is the mathematics of distortion or the mathematics of shape-change without adding to or subtracting from the original shape. An example is a convex mirror, such as found at the circus fun house. The reflection changes the image of a person without physically changing the person.

Let us explore the contributions of many people to Geometry.

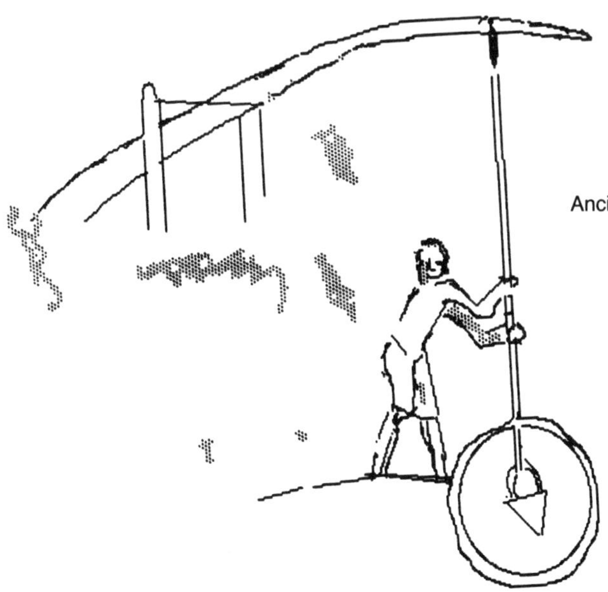

Ancient method of irrigation.

BABYLONIANS

The Babylonians had a system for recognizing many of the planets and their orbits as early as 3000 B.C. They measured circles and observed the stars as well. They knew another shape—the square. Babylon, the capital, was built on a square base. These early people knew how to build canals that harnessed the rich soil and water of the Tigris and Euphrates rivers.

One of their illustrious kings was Hammurabi who wrote a Code of Laws, which included land and business regulations, tariffs, wages, loans, and debts. This code also defined individual rights.

EGYPTIANS

The Egyptians (2900 B.C.) designed temples and pyramids with the use of plumb lines and triangles.

The Great Pyramid[2] has a base large enough to cover eight football fields. One hundred thousand slaves worked twenty or more years to build this wonder of the ancient world. Each limestone block weighed two and a half tons and needed a crew of forty men to move each block. The Great Pyramid is made up of over 2,300,000 blocks. Its height is 481 feet.

A pyramid.

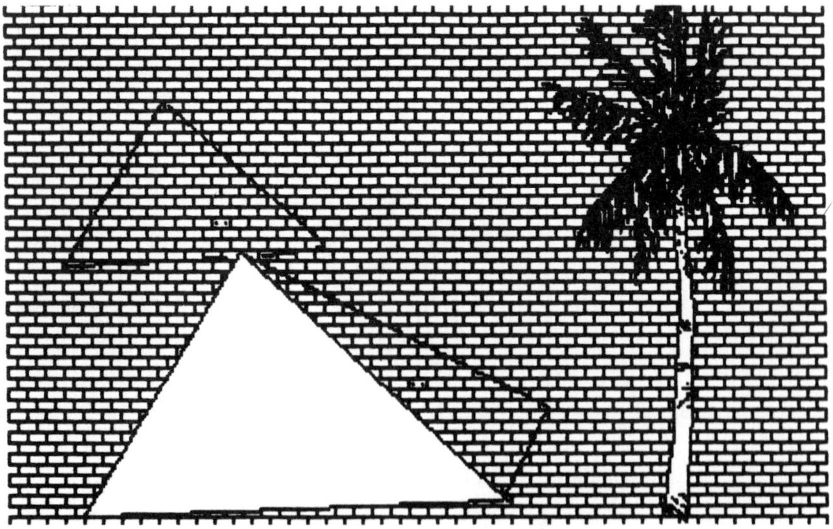

THALES (Greek)

Thales, a deep thinker and astronomer, started the study of angles, lines, and triangles (640–546 B.C.). The deductive character, which he gave to science, is his chief claim to mathematics distinction. Thinking deeply about mathematics, Thales had numerous accidents. The following is supposed to be a true tale:[3] Plato said that Thales fell into a well while looking at the stars. A pretty slave girl laughed at him, saying: "He wanted to know what happens in the heavens but he did not observe what was in front of his own feet!"

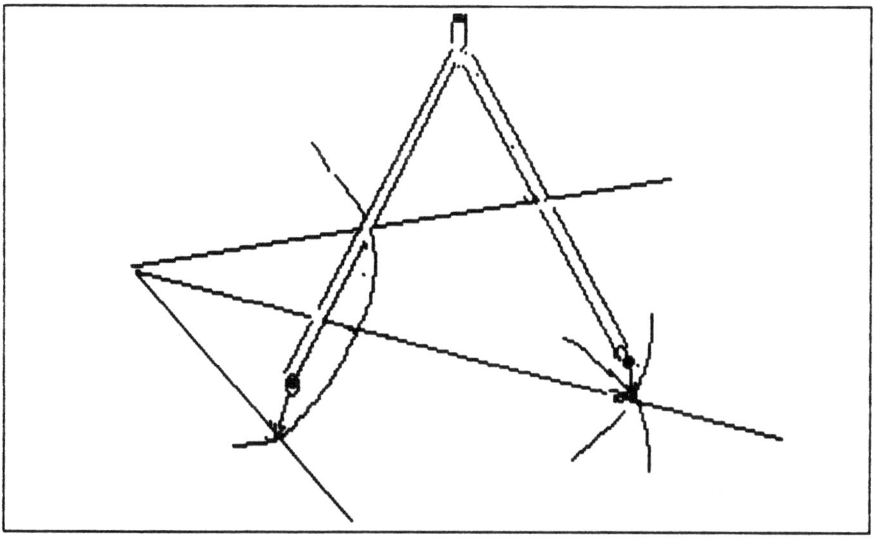

Angles, lines, and triangles.

PYTHAGORAS (Greek)

Remember this Greek philosopher? He gave us a famous theorem, which is: the square of the hypotenuse of a right triangle is equal to the sum of the squares of the other two sides. The theory, as you remember, repeats the joke, "The squaw on the hippopotamus is equal to the sum of the squaws on the other two hides."[4] Although the Babylonians had arrived at the same theorem about a thousand years before Pythagoras, it bears his name because Pythagoras (580 B.C.–?) proved the theory. It is a useful principle for carpenters to use so that rooms will be perfect rectangles.

The first important discoveries in Greek mathematics are ascribed to Pythagoras and his followers. Pythagoras also discovered the law of musical intervals, which denotes the mathematical relationship of the length of a musical string to the pitch of the note it produces.

Pythagoras taught that numbers were associated with all things, such as people, colors, and flowers.

PLATO (Greek)

Plato was a philosopher, not a mathematician. However, he is included in this text because of his impact on the development of mathematical theory.

Plato was curious about the elements of the earth and knew quite a bit about them. Today we know that all matter is composed of one or more parts. These parts are called "chemical elements." Some chemical elements known to the ancient world were: carbon, cobalt, gold, iron, lead, mercury, silver, and tin. The ancients thought of these substances as being composed of four components: fire, earth, air, and water. Plato, who lived from 427 B.C. to 347 B.C., added a fifth element, namely "world."

Today we know that each element not only has a name but also has an assigned symbol. Although there are more than a hundred elements, we know they differ from each other because of the distribution of their atomic proportions (electrons plus protons plus neutrons). Each element also has an atomic weight and may be combined with another to form a compound.

Plato's philosophy of elements played an important role in the development of mathematics. He believed that human material thoughts were temporary and that thoughts of geometric shapes were eternal. Plato also believed that the human soul could be purified by thinking of eternal qualities such as those found in geometry. Following is a chart showing the relationship of Plato's idea of elements with their geometric symbols:

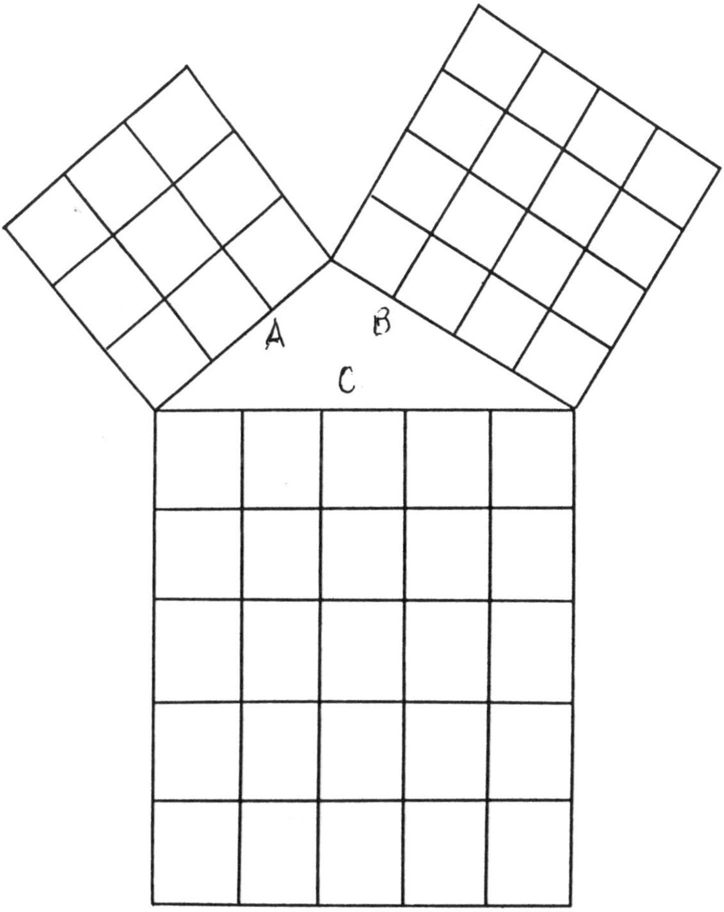

The Pythagorean Theorem: $A^2 + B^2 = C^2$. The square of the hypotenuse of a right triangle is equal to the sum of the squares of the other two sides.

Element	Solid Figure		Planes
Fire	Tetrahedron		4 Faces
Earth	Cube	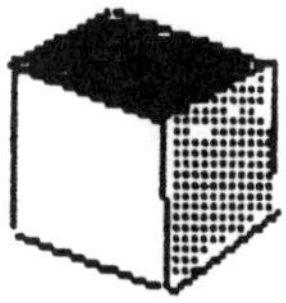	6 Faces.
Air	Octohedron		8 Faces
Water	dodecahedron	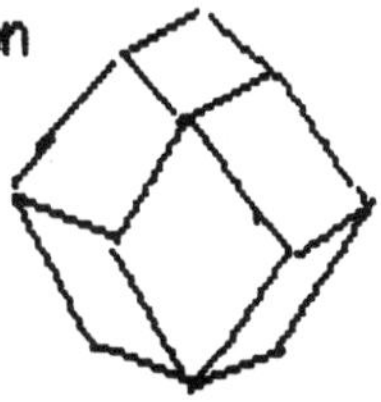	12 Faces
World	Icosahedron	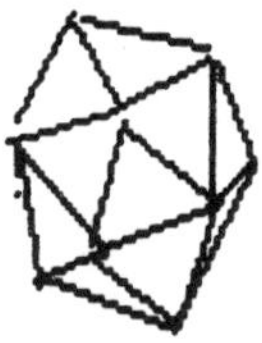	20 Faces

Plato's chart of elements.

Plato believed that the planets moved in a circle; therefore all circles and spheres were perfect shapes. He founded an academy near Athens which was continued for nine hundred years. Over the door the following motto was inscribed:

Let he who is ignorant of geometry not enter here.

ARISTOTLE (Greek)

All men are mortal
Socrates is a man
Therefore Socrates is mortal.
(*Syllogism: a form of reasoning.*)

Aristotle brought to mathematics his experiences as a psychologist, moralist, logician, biologist, political thinker, philosopher, and a talent for literary criticism. He developed the *axiom*, meaning "a statement accepted without question." It requires no proof. For example: "A whole is greater than its parts" or "If equals are added to equals, the sums are equal."

Aristotle lived from 384 B.C. to 322 B.C. He was concerned with mathematics and mathematical physics as supplying illustrations of correct mathematical reasoning. He used a systematic and detailed form of reasoning, which we call a syllogism (a statement of evidence followed by a conclusion).

EUCLID (Greek)

There is no royal road to geometry

—Euclid

Euclid (33 B.C.) systematized arithmetical and geometrical knowledge into a basic book, *The Elements*. This famous book consisted of thirteen volumes. He described and systematically proved every line, point, and shape in it. *The Elements* has lasted for over two thousand years. It is still studied today using Euclid's diagrams and proofs. Some of Euclid's postulates (self-evident truths) are:

Things equal to the same thing are equal.
All right angles are equal to each other.
A circle can be described with any point as center and with a radius equal to any finite straight line drawn from the center.

SYSTEMATIZED REASONING

If two sides of a triangle are equal, the angles opposite these sides are equal.

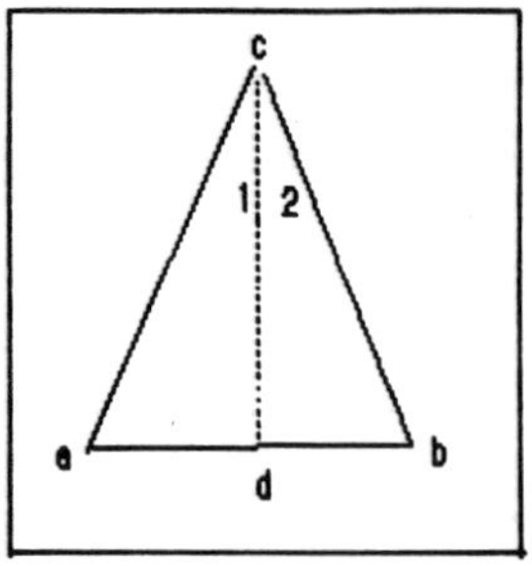

Given: Δ ACB, with CA = CB.
To prove: ∠ A = ∠ B.

Plan: To prove that ∠ A = ∠ B, we can show that ∠ A and ∠ B are corresponding angles of congruent triangles. Two such triangles, ACD and BCD, can be formed by drawing CD, the bisector (CD) of Δ ACB.

Proof: Statements	Reasons
1. Draw CD, the bisector of ∠ ACB.	1. Every angle has a bisector
2. ∠ 1 = L 2.	2. A bisector divides an angle into two equal parts.
3. CA = CB.	3. Given.
4. CD = CD.	4. Given.
5. ACD = BCD.	5. Given = s.a.s.
6. ∠ A = ∠ B.	6. Corresponding parts of congruent triangles are equal.

Galileo.

GALILEO (Italian)

This famous Italian was an astronomer, scientist, and philosopher as well as a mathematician. He lived from 1564 to 1642. Galileo gave us the law of the pendulum when he was twenty years old. He timed the swings of a lamp in a cathedral and noticed that each swing took the same time regardless of the size of the arc. Later he suggested that a simple pendulum could time the pulse rate of a medical patient. After experimenting he proved that every free-falling object falls to earth with a constant acceleration of thirty-two feet per second every second.

Newton's theories were based on Galileo's work with falling bodies. Galileo has been credited with making the first thermometer in 1593.

JOHANNES KEPLER (German)

Johannes Kepler (1571–1630), astronomer and mathematician, left us three laws of motion that were to guide Sir Isaac Newton later.

Kepler's Laws of Motion are:

1. Every planet follows its own path around the sun.
2. Planets move faster when closer to the sun.
3. The path around the sun is an ellipse, not a circle.

As a mathematician, Kepler worked with numbers that express relationships with other numbers. These numbers also simplify elaborate computations. They are called logarithms.

$4^3 = 4 \times 4 \times 4 = 64$
4 = base number
3 = the exponent that tells how many times 4 is to be multiplied by itself (expanded notation)
64 = the relationship of itself to 4

In other words, logarithms are numbers that quietly perform an operation as expressed by the base number and its exponent. In 1619, John Napier, a Scotsman, placed these numbers on a chart for easy reference. It is a "Table of Logarithms."

RENÉ DESCARTES (French)

We are indebted to Descartes (1596—1610) for using the first of the alphabet letters to represent known quantities and the last alphabetic letters to represent unknown quantities. He applied algebra to geometric measurement and made contributions to the theory of equations. This advanced geometry, using algebra to solve geometric problems, is sometimes called "Cartesian" geometry. Descartes, in his three-volume book, *Geometry*, Part I, systematically classified curves as circles, parabolas,[5] hyperbolas, and the ellipse. His measurement ideas are used today in the making of automobile headlight reflectors, projectiles, optical instruments like microscopes, eyeglasses, telescopes, and photographic lenses. Descartes was the first person to use a graph,[6] a study told by points and expressed by lines and bars.

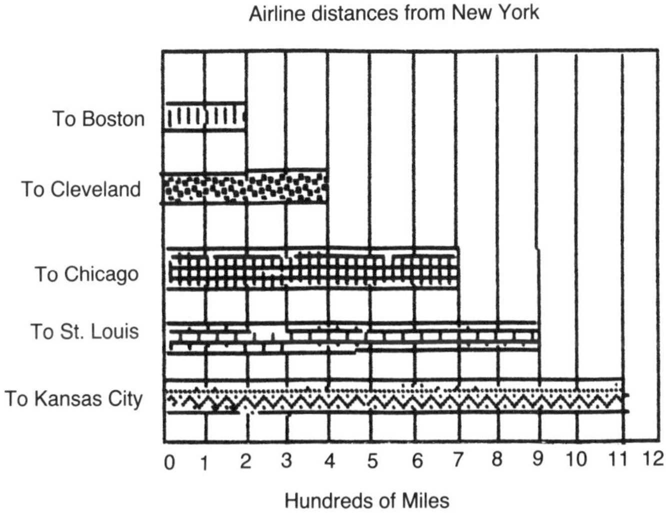

Example of a bar graph. Other graphs may be a line graph, a circle graph, or a pictograph.

LEONHARD EULER (Swiss)

This Swiss mathematician did much of his work when he became blind. He enjoyed composing formulas and is considered the founder of modern mathematical analysis. He studied the properties of geometric forms and learned that some geometric forms can be bent or stretched, but do not lose their original shape or form. For example, "a coffee cup is equivalent to a doughnut because one can be continuously deformed into the other, with the cup handle becoming the doughnut's hole. But both are different from a sphere because no amount of stretching creates or removes the hole."[7] This idea is called topology, the mathematics of distortion. Augustus Ferdinand Moebius

(1790–1868), a German astronomer and mathematician, first investigated the properties of topology. Euler built upon Moebius's idea.

Topology is different from the geometric study of angles and figures like triangles or squares. These plane figures remain unchanged in a rigid shape. The topological change of the shape is done by stretching or bending.

Euler was born in 1707 and died in 1788. He produced thirteen children and wrote 886 mathematics books and articles. Euler, like Benjamin Franklin enjoyed making "magic squares."

.08	.05	.14	?
.11	.17	.23	?
.20	.29	.26	?
?	?	?	?

Can you solve this square?

Steps for each row, column and diagonal

Sum of the end numbers			**Sums less the inside numbers**	
Row 1	.08	+ ,14 = ,22	,22	− ,05 = ,17
Row 2	,11	+ ,23 = ,—	—	− ,17 = ,17
Row 3	,20	+,— = ,46	,46	− ,29 = ,17
Column 1	,08	+ .20 = .28	.28	− .05 = .17
Column 2	—	+ .29 = .34	.34	−— = .17
Column 3	.14	+— = .48	.46	− .29 = .17
Diagonal	.08	+ .26 = ,—	.34	− .17 = 17
Diagonal	.14	+.— = .34	.—	− .12 = .17

Complete This Magic Square by filling in the missing numbers.

8		10	?
9	7		?
	11	6	?
?	?	?	?

The magic difference is ____.

Make Your Own Magic Square.

KARL FRIEDRICH GAUSS (German)

Karl Friedrich Gauss was one of the most important mathematicians of his time (1777–1855). As a child he was extremely gifted. His genius appeared before he was thee years old, when he mentally calculated an error in his father's payroll. At this time he taught himself to read and to calculate arithmetic. By age nineteen he proved that a polygon could have seventeen sides by using only a ruler and a compass. Gauss later developed the mathematics of electromagnetics. A unit of magnetism is called the "gauss" in his honor. Gauss also made measurements of curved surfaces. He introduced the concept of congruence (1801) and foresaw the importance of topology. Gauss invented and built a telegraph as well as an intercom system two years earlier than Samuel Morse.[8] He dominated nineteenth century mathematics.

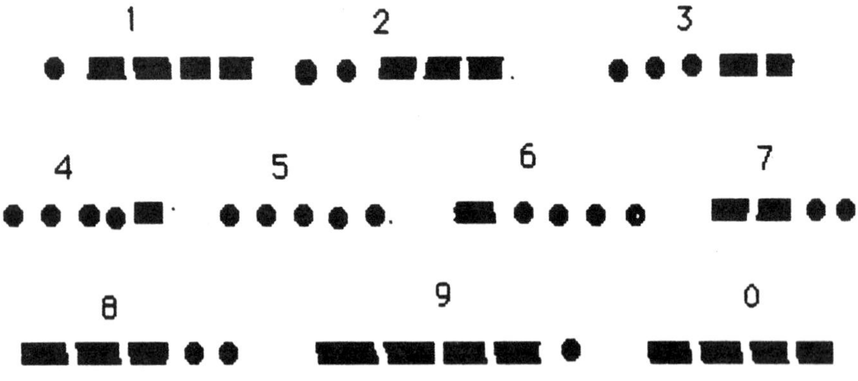

Numerals in international Morse code.

Albert Einstein.

ALBERT EINSTEIN (American)

Einstein revolutionized the world with new ideas of time, space, mass, motion, and gravitation. He gave us the idea of relativity, $e = mc^2$ (energy equals mass times the square of the speed of light). Two postulates are an outgrowth of Einstein's mathematical theory.

These are: the relativity principle is true for all phenomena. The second postulate deals with the velocity of electromagnetic waves in empty space. The mass of an object increases with its velocity. Interchange between mass and energy is when energy is liberated from the nucleus of the uranium atom. Some of the mass is transformed to energy. Mass is not unchangeable. The discovery of $e = mc^2$ led to the introduction of nuclear physics. This theory also led to the "electric eye" and to the invention of television.

Recently French astronomers have found rings in space that Einstein had predicted. Using spectroscopic analysis, these scientists have proved that the Einstein ring is nothing but light bent around an enormous mass in the shape of a ring. The discovery makes it possible to find other galaxies in the universe by determining their spectroscopic effect on light.

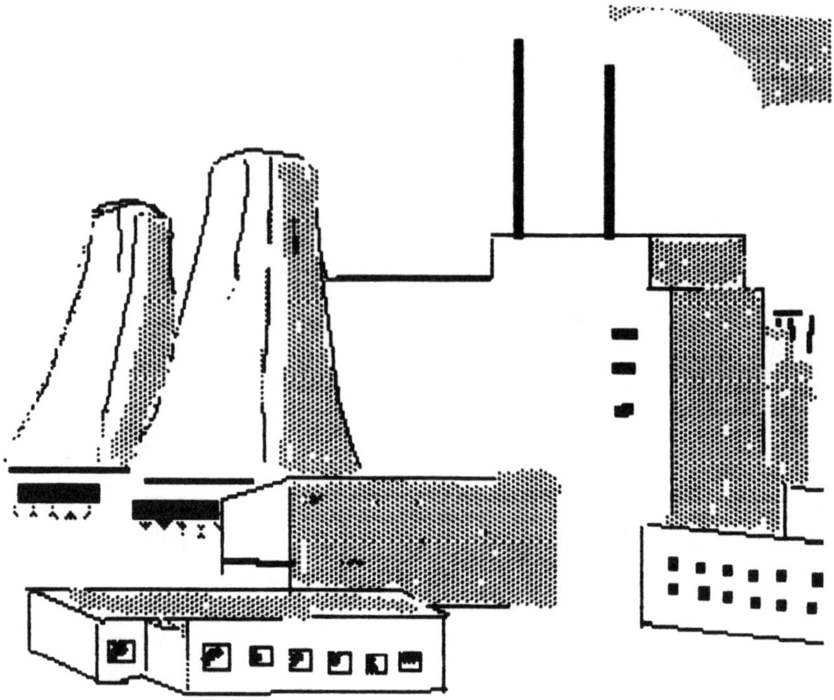

An atomic reactor plant. These reactor plants generate electricity.

Notes

1. Anthony Revelli, *An Adventure in Geometry* (New York: Viking Press, 1957).
2. *World Book*, 1963 ed., s.v. "The Great Pyramid."
3. B. L. van der Warden, *Science Awakening* (New York: Oxford University Press, 1963), 86.
4. D. Wells, *The Penguin Dictionary of Curious and Interesting Numbers* (London: Penguin Books, 1951), 89.
5. Apollonius, a Greek mathematician, studied the properties of parabolas as early as 220 B.C. However, results obtained by the analytic method of Descartes are used today because it is easier.
6. C. B. Upton and K. G. Fuller, *American Arithmetic, Book 6* (New York: American Book Co., 1960), 28.
7. J. Gleick, "One of Math's Major Problems Reported Solved," *The New York Times*, 30 Sept. 1986, p. 19f.
8. Morse's telegraph improved earlier inventions with a relay device, the dot-dash code, and an improved electromagnet.

5

TRIGONOMETRY

Trigonometry can measure heights or distances by using the relationship of the sides and angles of a triangle. Trigonometry can also measure the pattern of radio waves by using the relationship of cycles and angles.

HIPPARCHUS (Greek)

Hipparchus, who lived from 160 B.C. to 125 B.C., was the founder of plane trigonometry and used it to distinguish between the different lengths of the seasons. He also made the first catalog of star magnitudes[1] (star brightness) and then marked their position on a globe. Hipparchus knew the distance of the moon from the earth. He was the first person to point out places on earth via latitude and longitude.

PTOLEMY (Greek)

Another deserving mathematician and astronomer was Claudius Ptolemy (100–170). Ptolemy wrote *The Amalgest*, in which he attempted to prove that the earth was the center of the universe. This idea dominated scientific thought for fourteen hundred years. In his book he used degrees, minutes, and seconds in the measurement of angles. Ptolemy noted the spherical shape of the earth, gave the value of π, determined the positions of 1,028 stars and knew five of the planets. A book on *Sound* is attributed to him. He also made maps based on his mathematical calculations. Much of Ptolemy's work was based on the findings of Hipparchus.

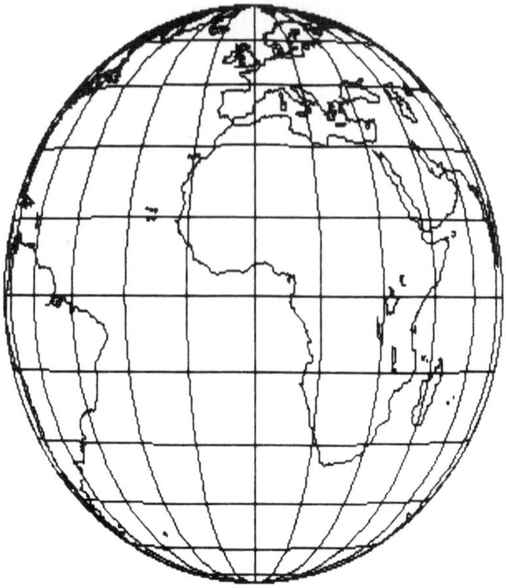

Latitude and longitude. Latitude runs east and west and is measured by degrees north and south of the equator. Longitude runs north and south and is measured by degrees east and west of the meridian of Greenwich.

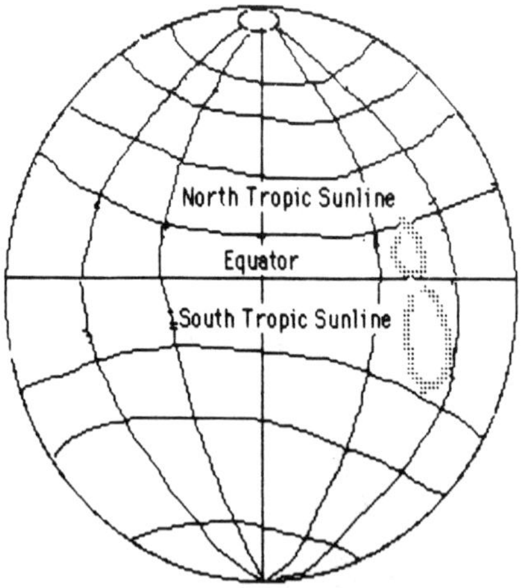

The analemma. We can tell the sun's tropical position for each day. On a chart, each position is pinpointed daily by a dot. Direct rays move directly across the equator to the tropic sun lines a day at a time. By year's end the imaginary dotted figure 8, called the analemma, will have been formed. It is measured in degrees, minutes, and seconds, too.

JOHN NAPIER (Scottish)

In a period from 1559 to 1617, Napier was an early advocate of the decimal system, which is based on the number 10. With the symbols of base 10, Napier made a set of multiplication tables in the form of sticks, which could also divide and extract square root.

In addition to developing the decimal system, Napier eliminated the drudgery of having to multiply large numbers by making a table of reference, which was called a "Table of Logarithms." This table is essentially used today by scientists and engineers.

The following table explains to what power a number must be raised in order to produce another equal number. This table is based on powers of ten:

10^1 = 10—the logarithm of 10 is 1
10^2 = 100—the logarithm of 100 is 2
10^3 = 1000—the logarithm of 1000 is 3
10^4 = 10,000—the logarithm of 10,000 is 3

Napier, a serious scientist, was considered to be quite eccentric. There are many anecdotes about him. Once he was extremely irritated because his neighbor's birds flew continually over his land. Napier repeatedly complained to the neighbor, who refused to cooperate. The neighbor told Napier that if he could keep the birds from flying over his land, he could have them. Napier soaked some corn in whiskey and strew the soaked corn over the land and waited. The birds ate the corn and began to wobble. Napier collected the sick birds and disposed of them.[2]

BENJAMIN BANNEKER (American)

Benjamin Banneker[3] (1731–1806), a black mathematician, made surveying plans for our nation's capital, Washington, D.C. after Pierre Charles l'Enfant withdrew in a huff at Congress. Banneker had memorized the data and finished the project, much

Benjamin Banneker. One of the surveyors of Washington, D.C. (1791).

to the delight of Thomas Jefferson. Although he had little formal education, Banneker studied at night after working all day as a farmer. He taught himself mathematics and astronomy and was tapped by Jefferson to send one of his almanacs to the French Academy of Sciences. Banneker's almanac contained weather predictions, tide tables, and farmers' tips. Banneker is also known for his wooden clock, which kept accurate time and chimed on the hour.

Notes

1. Ptolemy was the first person to list stars according to magnitude in A.D. 150.
2. H. Eves, *An Introduction to the History of Mathematics* (Philadelphia: Saunders College Publishing, 1983).
3. G. Reasons and G. Patrick, *They Had a Dream* (Los Angeles: Los Angeles Times Syndicated Square, 1969), 12.

6

CALCULUS

Calculus is usually learned after the mastery of algebra, geometry, and trigonometry. It deals with extremely small changes in the rate of speed or changes in quantities. Calculus is used primarily in physics, engineering, and aerodynamics.

The discovery of calculus helped to mark the beginning of modern mathemetics. Archimedes (287–212 B.C.), Newton (1642—1727), and Von Leibnitz (1646–1716) uncovered the same basic ideas of calculus working independently of each other. Each receives the same credit.

ARCHIMEDES (Sicilian)

Archimedes (287–212 B.C.) gave us formulas for finding the volume and surface area of a sphere. He also identified a spiral as a shape of motion that moves around a fixed point constantly, either receding or approaching. He calculated the number of

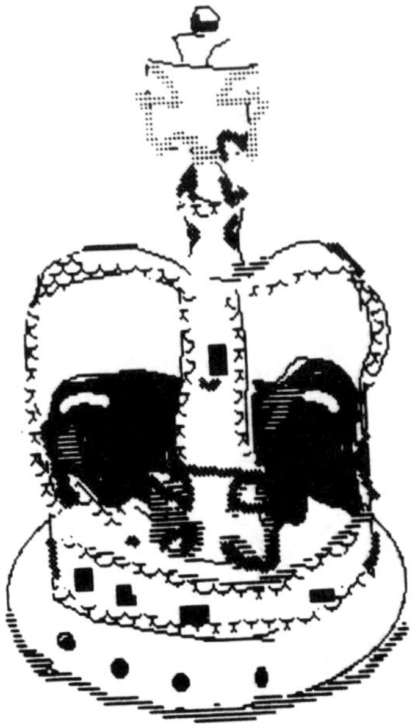

A crown. This symbol of power tells how calculus helped a king.

grains of sand which could fill the universe. Archimedes discovered that a body displaces an amount of water equal to its own volume. In addition, Archimedes found that salt water and fresh water have different densities.

He could work on problems for long periods, never stopping to eat or sometimes, bathe. There are many stories about this famous mathematician. One story concerns his solving the problem of the king's crown. Gittleman[1] gives the following version of this story:

> The king ordered a crown of gold but was afraid the jeweler had mixed in cheaper material, so he asked Archimedes to check to see if the crown were made of pure gold. One day, while in the tub, an idea came to him. Archimedes suddenly leaped from the tub and ran down the street naked, shouting "Eureka, Eureka, I have found it!"
>
> The plan he decided to use was to immerse the crown in water, along with equal weights of gold and silver, and then measure the water displaced. If the crown were pure gold, it should have displaced the same amount of water. The crown was not pure gold.

SIR ISAAC NEWTON (English)

Sir Isaac Newton[2] (1642–1727) is considered a co-founder of calculus, which analyzes mathematically all movement and all change. From this analysis, Newton arrived at the laws of motion and gravity. Some of these laws are:

1. An object at rest will remain at rest, unless acted upon by an external force.
2. All objects in the universe exert gravitational attraction upon each other.
3. Every action has an equal and opposite reaction.

You remember the famous apple story. Newton was drinking tea in the garden when he saw an apple fall. He realized that the force that kept the moon in orbit was the same force that pulled the apple to fall to the earth. His theory was offered in 1666 and was not fully accepted until 1686. In 1687 he published his findings in *Naturalis Principia Mathematica* upon the urging of Edmund Halley, the English astronomer.

Sir Isaac Newton had a reputation for deep concentration. It is said that he was riding home one day and dismounted to let his horse walk up a steep hill. At the top of the hill, he turned to mount his horse only to discover that the horse had run away and he was left with an empty bridle in his hand.

GOTTFRIED WILHELM VON LEIBNITZ (German)

Gottfried Wilhelm von Leibnitz (1646–1716) invented a calculating machine and shares the distinction of developing calculus with Sir Isaac Newton. He is also given credit as the inventor of the binary system, which counts only from zero to one and then starts a new place. Binary numbers use more places but the system is easier to translate into computer electronic impulses.

Decimal System	Binary System
1	1
2	10
3	11
4	100
5	101
6	110
7	111
8	1000
9	1001
10	1010

SOPHIE GERMAIN (French)

Sophie Germain used the pseudonym "Monsieur le Blanc" because women[3] in her day (1776–1831) were highly criticized for daring to delve into "man's work." She is called the founder of mathematical physics.

Mme. Germain taught herself Latin in order to read the works of Leonhard Euler and Sir Isaac Newton. She had to take mathematical correspondence courses from L'Ecole Polytechnique which did not accept women. Her interests lay in the

The Eiffel Tower, a huge skeletal wrought-iron tower in Paris that Gustave Eiffel designed for the Paris Exposition of 1889. This wonder of architecture contains more than seven thousand tons of iron and steel.

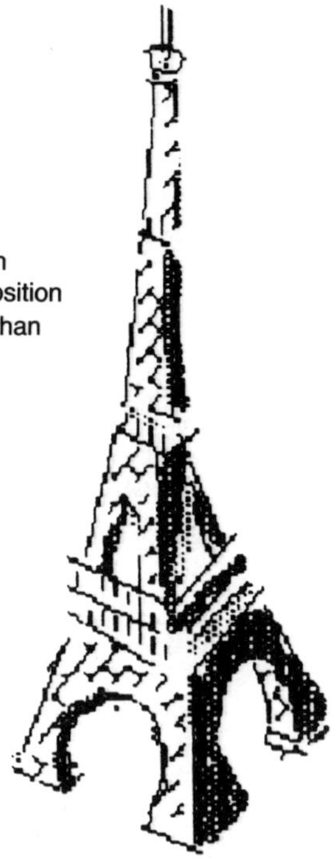

theory of numbers, sound, and elasticity. She exchanged letters with leading mathematicians of the day under her nom de plume. The Institut de France offered a prize to anyone who could explain the phenomenon of vibrating metal plates. Sophie Germain presented three memoirs on the subject and won the prize. Subsequently, she established an equation for the elasticity of vibrating metal surfaces. Her work helped architects to design specifications for the construction of the Eiffel Tower.

Notes

1. A Gittleman, *History of Mathematics* (Columbus, Ohio: Charles E. Merril, 1975), 67.
2. T. Glazer and D. Evans, "A Ballad of Sir Isaac Newton," *Space Songs* (New York: Argosy Music Corp., 1959).
3. Other female contributors to mathematics are: Hyapatia (Greek), who was also distinguished in medicine and philosophy; Maria Agnes (Italian); Sonya Kovalevskaya (Russian); Marie Curie (French); and Grace Hooper (American).

7

PROBABILITY

Probability is the expected mathematical outcome of an event. It may be a number that tells how likely a certain event will happen, like betting on a horse at the races or setting insurance rates. Probability, in other words, is the frequency with which one phenomenon takes place in relation to possible alternatives. Tossing a coin to determine "heads" or "tails" is another example.

BLAISE PASCAL (French)

Blaise Pascal was a religious philosopher, a monk, and a mathematical genius (1623–1662). Before he was sixteen he had completed *The Geometry of Conics*.

With Pierre de Fermat, Pascal is credited with co-authoring the theory of probability. The idea came about when a friend wanted to win gambling bets and asked Pascal for a more precise method of determining the odds. So he thought about the idea for two years. Then a plan was decided upon and the theory of probability was born. Later, Pascal became fascinated with the geometric figures that shadows made. From this point, his work on conic figures began in earnest.

At age eighteen, Pascal invented an adding machine that could also subtract. Pascal is credited with the invention of the one-wheeled wheelbarrow. He also developed the science of hydramatics (pressurized fluids in motion). Wherever hydraulic pressure is needed, his basic idea is used today—for pistons, autos, caissons, jacks, cranes, and elevators.

8

STATISTICS

Probability and statistics help schedule the work of factories and find the best route for the collection of trash and garbage. The study of statistics began in the 1700s. It is the science of collecting, classifying, analyzing, and interpreting the resulting numerical facts. Insurance companies make great use of comparing numbers.

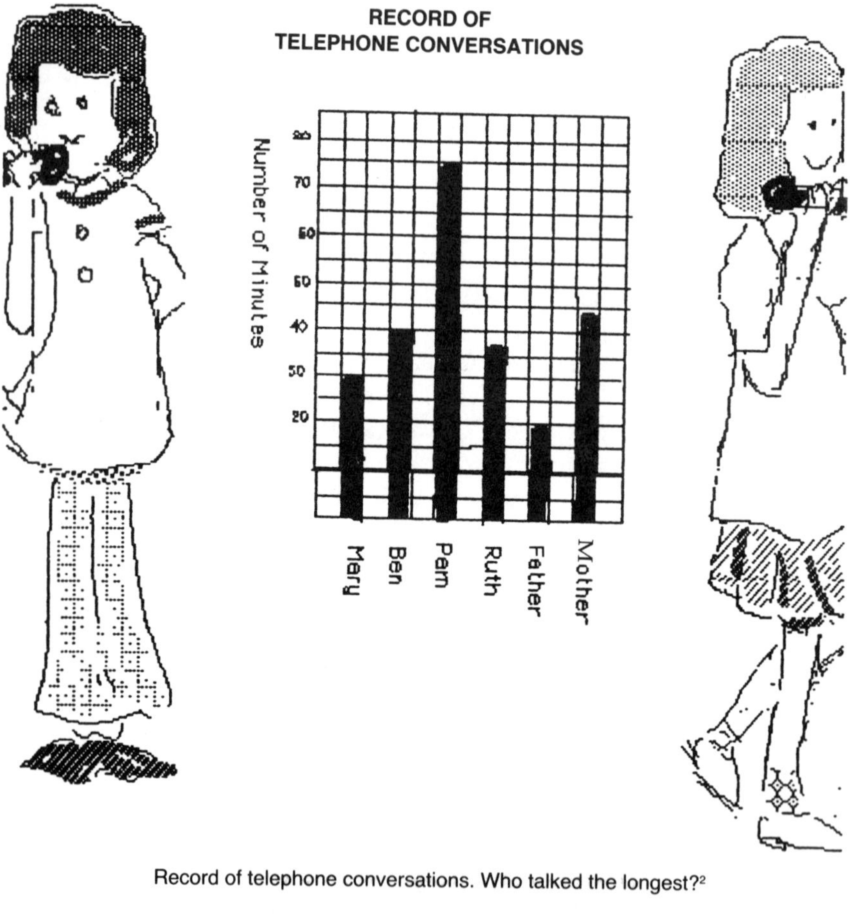

Record of telephone conversations. Who talked the longest?[2]

KARL PEARSON (English)

Karl Pearson (1857—1936), a professor of mathematics, is known as the founder of statistics. He first applied mathematical methods to the study of eugenics. Pearson wanted to improve the quality of the human race by genetically bettering offspring. Today scientists use computers to help search for the best way to study life.[1] This quest offers a simulation of biological functions from plant growth to animal processes and is called artificial life.

Statistics began in Germany, where it was used to tax citizens and count the population. After counting, the resulting numbers were interpreted and conclusions were drawn from these facts. Today most businesses and industries depend upon the findings to determine rates. Statistics can be applied to most information-gathering groups, including public polls. The use of statistics has become widespread. Probability is closely entwined with statistics.

Notes

1. James Gleick, *The Ann Arbor News*, 5 November 1987, p. 6A.
2. Random House, *The Random House Mathematics Program*, 2nd. ed., book 5 (New York: Random House, 1975), 218.

9

COMPUTERS

In recent years, mathematicians have been solving problems via computers. Computers are electronic devices that can routinely solve problems at high speeds. Some computers can store billions of characters and operate millions of times faster than a person.

Some of the uses of the computer are: predicting earthquakes, creating greeting cards, compiling concordances, diagnosing health ills, playing games, computing taxes, transacting bank business, library check-out systems and composing music, teaching foreign languages, designing clothing and hair styles.

These welcome servants operate essentially as a digital or as an analog computer. The digital operates by the use of counting, i.e., quantities are expressed as numbers. Measurement is the main characteristic of the analog, i.e., quantities are expressed as voltage and the like. Many computers are considered "hybrid" because they have features of both the digital and the analog.

COMPUTER DEVELOPMENT

Between 1975 and 1980, computer knowledge technology changed and has been greatly improved. In fact, the computer is now a status symbol among many families. Since there have been numerous contributors to the success of this machine, I shall mention only a few of these personalities.

Blaise Pascal (1623—1662), a Frenchman. In order to help his father collect taxes, Pascal invented the first digital adding machine in 1642. This machine could add and subtract. Pascal's basic ideas are still used today in automobiles and other machines. For instance, the speedometer indicates the speed of an automobile. The hydraulic jack, airplanes, elevators, and hydraulic presses are a few examples of how moving pressurized liquids serve mankind.

In 1801, a French engineer, **Joseph-Marie Jacquard** (1752–1834), invented an automatic loom. It used a chain of punched cards, which was set to rotate and pass through the loom needles, but only where the needles matched the holes. The threads determined the pattern. In the United States, Jacquard's machine technique was also used to handle census figures.

In 1835, **Charles Babbage**, an English mathematician, proposed the first automatic digital computer; however, he died before it was finished. **Howard Aiken**, a Harvard Mathematician, completed the task in 1944.

Vannevar Bush (1890–1974), an American electrical engineer, built the first analog computer in 1931. It was faster than the digital computer, but not as accurate or dependable. The analog computer was built to interpret and translate problems and store the rest into symbols. This early analog computer was a collection of gears, gauges, and wires. Today, a modified type of analog computer would be bathroom scales. The scale numbers show the amount of weight. Translating this number into pounds indicates how heavy the object is.

The American, **Norbert Wiener** (1894–1964), in designing high speed electronic computers, coined the word "cybernetics." Cybernetics deals with mechanical reactions to electrical demands. Today, for instance, the roof of a convertible automobile raises its protective roof when it "feels" rain. An airplane "responds" to the automatic pushbutton set by the pilot. Wiener graduated from Tufts at age fourteen and earned a Ph.D. from Harvard at age eighteen.

From Hungary, **John von Neuman** (1903—1957), a Princeton mathematician, made studies of the brain, which were helpful to him in storing information into the digital computer. He possessed a photographic mind yet on occasion could not find drinking glasses in his home. He sometimes became so immersed in mathematics that he had to call his office to find out why he had taken a trip.

At present these computers are advanced in storage space, quick to respond, and dependable in memory. Today, most monitors are in color and some can now respond to the voice rather than answer to a typed command. Some computers use laser printers.

Grace Hopper (1904–), an admiral of the U.S. Navy, aided the fast-paced growth of the computer by developing an electric "dictionary," an improvement over machine codes, which developed into a telemetry system and was first used in the search for the *Titanic.* She also gave us the first machine translation system called "Autocade" (today called Math Matic). In the middle of the 1950s Grace Hopper developed a machine code compiler that could understand instructions not only in English but in French and German, as well. We are exploring Artificial Intelligence as an outcome of Grace Hopper's gifts.

In the early days, machine languages in binary code were used to program computers. Wide use of computers during a later period was facilitated by the introduction of high-level programming languages. Languages resembling natural language are now interpreted or compiled into the binary code, which is the "native" language of the computer. It was during this period that the home computer, as well as the personal or desk-top computer, became popular. This advance did much to modify a way of doing things, whether just balancing the family checkbook or keeping a record of Christmas cards. The various computers of the Apple II line became very familiar in the classroom throughout the country. A new generation of desk-top computers was introduced by Apple in January 1984 with the introduction of the Apple Macintosh. There are numerous other brands, such as IBM, Digital, Commodore, Atari, Amiga, all of which have made their contributions to this "revolution" —telecommunications, laser printing, and electronic mail.

From counting on fingers and toes to sending electric messages, man and machine together have made mathematical history.

Yes, mathematics is truly a universal language. Here's hoping that you enjoyed a peep into the lives of some worldwide mathematics personalities and that you will remember how indebted we are to their genius and imaginative dedication throughout the ages.

BIBLIOGRAPHY

Adler, Irving. *Giant Golden Book of Math.* New York: Golden Press, 1960.

Ball, W. W. Rouse. *A Short Account of the Story of Mathematics.* 4th ed. New York: Dover, 1960.

Bergamini, David. *Mathematics.* New York: Life Science Library, 1968.

Bocke, Kees. *Cosmic View.* New York: John Day & Co., 1955.

Boyer, Carl B. *A History of Mathematics.* Princeton, N.J.: Princeton University Press, 1985.

Davidson, Basil. *The Lost Cities of Africa.* New York: Little Brown, 1959.

Davis, Gordon. *Computer Data Processing.* New York: McGraw-Hill, 1973.

Diop, Cheika Anta. *L'afrique noire précoloniale.* Paris: Présence Africaine, 1960.

Durawa, Alexander. "Getting Started in Personal Computers," *Academic American Encyclopedia* (Electronic Edition), 1984.

EDUCOM Computers for Humanists. "Machines as Helpers." *Bulletin of the University Communications Council.*

Eves, Howard. *An Introduction to the History of Mathematics.* Saunders College Series. Philadelphia: Saunders Publishing, 1983.

Gittleman, Arthur. *History of Mathematics.* Columbus, OH: Charles E. Merrill, 1975.

Glazer, Tom and D. Evans. "A Ballad of Sir Isaac Newton" in *Space Songs,* Side 2. Motivation Records. New York: Argosy Music Corp., 1959.

Glazer, Tom and D. Evans. "Constellation Jig" in *Space Songs,* Side 1. Motivation Records. New York: Argosy Music Corp., 1959.

Gleick, James. "One of Math's Major Problems Reported Solved." *The New York Times,* September 30, 1986.

Gleick, James. "Computers Aid in Quest for Elusive Secrets of Life." *The Ann Arbor News*, Thursday, November 5, 1987, section D1.

Muller, Frances. *Arithmetic*. Englewood Cliffs, NJ: Prentice Hall, 1956.

Osen, Lynn. *Women in Mathematics*. Cambridge, MA.: MIT Press, 1951.

Random House School Division. *The Random House Mathematics Program*. New York: Random House, Inc., 1974.

Ravelli, *Anthony. An Adventure in Geometry*. New York: Viking Press, 1957.

Reasons, George and Sam Patrick. "They Had a Dream." Los Angeles: Los Angeles Times Syndicate, Times Mirror Square, 1969, p. 12.

Reid, Constance. *From Zero to Infinity*. New York: Thomas Crowell, 1955.

Rogers, James. *Story of Math for Young People*. New York: Pantheon, 1966.

"The Great Pyramid." *World Book*, Volume 14.

Upton, Clifford and Kenneth G. Fuller. *American Arithmetic*, Book 6. New York: American Book Co., 1960.

Waden, B. L. van der. *Awakening*. New York: Oxford Univ. Press, 1963.

Wells, David. *The Penguin Dictionary of Curious and Interesting Numbers*. London: Penguin, 1987.

Woodson, Carter G. *The African Background Outline*. New York: New American Library, 1969.

Zaskavskia, Claudia. *Africa Counts*. Boston: Prindle, Weber and Schmidt, 1969.

APPENDIX. Suggested Teacher/ Student Activities

1. Mathematics Tools Exhibit
2. Mathematics Personality Parade (Costumed)
3. Time Line
4. T.V. Show: Women in Mathematics
5. Map Project: Location of Cities, Countries via Latitude, Longitude
6. Puppet Show: Personalities
7. Insurance Rates Discoveries
8. Geometric Figures Display
9. Original Magic Squares
10. Computer Advancements
11. Architecture Throughout the Years
12. Selection of Mathematicians (not in textbook)
13. Games of the World
14. Styles of the Times
15. Climates of Birthplaces
16. Problems of the Times
17. Evolvement of Mathematical Functions Signs
18. Original Problems
19. Probability Games
20. Letters to Grace Hopper
21. Original Songs and Dances
22. Radio Program: Mathematics
23. Keeping a Budget
24. Bank Services
25. The Stock Market

INDEX